Introduction to Logic

ALSO BY WILLIAM R. PARKS

Paperback editions

Algebra

Beginning Algebra

Computer Mathematics Series

Sets Numbers and Flowcharts

Computer Number Bases

Computer Number Bases – Printed Text with Audio CD

Introduction to Logic

Boolean Algebra and Switching Circuits

Artificial Intelligence Series

Introduction to Artificial Intelligence

Statistics

Introduction to Statistics

Introduction to Gambling Theory – Know the Odds

Introduction to Logic

William R. Parks, B.S., M.S., Ed.M.

Hershey Books
www.wrparks.com

ISBN: 978-0-88493-048-8

INTRODUCTION

This text covers the fundamental concepts of logic used in human reasoning and especially in modern technological fields such as computer science, business data processing and computer applications. The material is presented in a simple and easy to understand format. We utilize a modified form of programmed instruction (PI) in this presentation. Small bits of information are presented in each frame before advancing to the next frame. Exercises are listed after several frames, followed by an answer key.

We also use example problems with an immediate presentation of the solution given right after the problem.
The immediate answer feedback approach to a problem is important when learning new concepts. The exercises serve the same purpose. Answers are given right after each set of exercises.

The study of logic goes hand and hand with the study of computer programming and basic concepts of flowcharting. The mathematics of set theory is also related to logic and is used to explain the basic principles of flowcharting for scientific and business data processing applications while observing basic rules of logic. Because of these interrelationships, we encourage the reader to acquire the other titles in this computer math series mentioned in the first page of this text.

INTRODUCTION TO LOGIC

1. In this study we will be concerned with symbolic logic. In symbolic logic letter symbols are used to represent declarative sentences.

EXAMPLES:

p: John is playing the violin.

q: Mary is shutting the door.

\- - - - - - - - - - - -

Note: The letter "p" stands for the entire sentence, "John is playing the violin." The letter "q" stands for the entire sentence, "Mary is shutting the door."

2. Symbolic logic is concerned with statements which are either true or false, but not both.

EXAMPLES:

(a) The White House is in Washington, D.C.	TRUE
(b) Chicago is in Texas.	FALSE
(c) What is your name?	NO TRUTH VALUE
(d) Please listen to me.	NO TRUTH VALUE

\- - - - - - - - - - - -

Note: It is true that the White House is located in Washington, D.C., therefore, we assign a truth value of "TRUE" to the sentence, "The White House is in Washington, D.C. The "TRUTH VALUE" of a sentence is either "true" or "false" but not both. "Truth value" is the "logical value" of a sentence which is either "true" or "false."

3. Two simple statements can be combined to form a compound statement. The following connectives are used between simple statements:

and : $\wedge$	conjunction
or : $\vee$	disjunction

EXAMPLES:

(a) John is playing the violin and Mary is shutting the door.
$p \wedge q$

(b) John is playing the violin or Mary is shutting the door.
$p \vee q$

- - - - - - - - - - - -

Note: The logical "and" in above Frame 3 looks like an upside down "V" shaped symbol. The term "conjunction" associated with this symbol refers to joining two statements with the "and" connective. The logical "or" looks like a "V" shaped symbol. The term disjunction associated with this symbol refers to joining two statements with the "or"connective.

4. The following connectives are also frequently used in symbolic logic:

not : $\sim$	negation	(refers to denial)
if-then : $\longrightarrow$	conditional	(refers to implication)

EXAMPLES:

(a) p : John is playing the violin.
$\sim p$: John is not playing the violin.

(b) p : John is playing the violin.
q : Mary is shutting the door.
$p \longrightarrow q$: If John is playing the violin, then Mary is shutting the door. (p is the hypothesis and q is the conclusion.)

- - - - - - - - - - - -

EXERCISE 1

Multiple choice questions:

1. Which symbols stand for sentences in symbolic logic.
(a) numerals (b) letters (c) words (d) none of these

2. Symbolic logic is concerned with statements that are either true or false.
(a) true (b) false

3. Which statement type uses the "and" connective.
(a) conditional (b) negation (c) conjunction (d) disjunction

4. Which statement type uses the "or" connective.
(a) conditional (b) negation (c) conjunction (d) disjunction

5. Which symbol stands for the if-then conditional.
(a) v (b) * (c) the arrow (d) none of these

6. Which is the negation of statement "p."
(a) not p (b) <P> (c) q (d) none of these

7. The compound statement "John is running and Mary is walking," can be represented by:
(a) $p \vee q$ (b) $p \wedge q$ (c) $p \rightarrow q$ (d) none of these

8. In symbolic logic a statement can be both true and false.
(a) true (b) false

9. In symbolic logic the statement $p \rightarrow q$ is called:
(a) a negation (b) a conjunction (c) a disjunction (d) a conditional

10. The negation symbol in symbolic logic is:
(a) $\rightarrow$ (b) $\wedge$ (c) ~ (d) *

ANSWERS FOR EXERCISE 1

1. (b) letters

2. (a) true

3. (c) conjunction\

4. (d) disjunction

5. (c) the arrow

6. (a) not p

7. (b) $p \wedge q$

8. (b) false

9. (d) conditional

10. (c) ~

5. We can use p and q as previously defined and write compound statements which contain more than one connective.

EXAMPLES:

(a) $p \wedge (\sim q)$: John is playing the violin and Mary is not shutting the door.

(b) $(\sim q) \longrightarrow (\sim p)$: If Mary is not shutting the door, then John is not playing the violin.

(c) $[(p \longrightarrow q) \wedge p] \longrightarrow q$: If John plays the violin, then Mary shuts the door. John is playing the violin. Therefore, Mary is shutting the door.

- - - - - - - - - - - -

Note: In the above examples we connect statements p and q with the following connectives -- also called logical operators:

Negation (not): $\sim$

Conjunction (and): $\wedge$

Disjunction (or): $\vee$

Implication (if...then): $\rightarrow$

"And," (if ...then) and "or" are called binary connectives because they join two sentences to make a compound statement. However, negation is a unary connective. It applies to a single statement and does not connect two statements.

- - - - - - - - - - - -

EXERCISE 2

Let p and q stand for the same two sentences about John and Mary. Write the sentences in English for the following logical statements:

1. $(\sim p) \rightarrow (\sim q)$

2. $[\,(p \rightarrow q) \wedge (\sim p)\,] \rightarrow (\sim q)$

3. $(\sim q) \rightarrow (\sim p)$

4. $p \wedge (\sim q)$

5. $\sim p$

6. q

ANSWERS FOR EXERCISE 2

1. If John is not playing the violin, then Mary is not shutting the door.

2 If John is playing the violin, then Mary is shutting the door. John is not playing the violin. Therefore, Mary is not shutting the door.

3. If Mary is not shutting the door, then John is not playing the violin.

4. John is playing the violin and Mary is not shutting the door.

5. John is not playing the violin.

6. Mary is shutting the door.

6. Truth tables are used in symbolic logic to define the meaning of connectives. In the following truth table:
p: I washed the windows.

(a) negation

p	~p
T	F
F	T

(I did not wash the windows.)

Note: In the above truth table if p is true: (I washed the windows), then (~ p is false), that is: "I did not wash the windows" is false. However, when p is false then not p is true: (~p is true).

(b) conjunction

p	q	$p \wedge q$
T	T	T
T	F	F
F	T	F
F	F	F

(I washed the windows and cleaned the floor.)

RULE: The statement $p \wedge q$ is true only if both p and q are true.

Note: A compound statement in logic is built from simple statements using the logical connectives. The truth value of p and q written as ($p \wedge q$) depends on the individual truth values of p and q.

(c) disjunction

p	q	$p \vee q$
T	T	T
T	F	T
F	T	T
F	F	F

(I washed the windows or cleaned the floor.)

RULE: The statement $p \vee q$ is true if at least one of the two statements is true.

(d) conditional

p	q	$p \rightarrow q$
T	T	T
T	F	F
F	T	T
F	F	T

(If I washed the windows, then I cleaned the floor.)

RULE: The statement $p \rightarrow q$ is false only when p is true and q is false.

EXERCISE 3

True or False questions:

1. The statement $p \wedge q$ is true only if both p and q are true.

2. If $\sim p$ is true then p is false.

3. If p is true and q is false then $p \vee q$ is false.

4. If p is true and q is false then the statement $p \rightarrow q$ is false

5. For two statements p and q there are four different possible combinations for conjunction, disjunction and conditionals.

- - - - - - - - - - - -

ANSWERS FOR EXERCISE 3

1. True

2. True

3. False

4. True

5. True

7. Another connective that we use in symbolic logic is the biconditional. It is symbolized by ⟷ . It connects two statements which are equivalent. Two statements are equivalent if they are either both true or both false.

EXAMPLE:

p	q	p ⟷ q
T	T	T
T	F	F
F	T	F
F	F	T

(The game will be called off if and only if the players don't show up.)

- - - - - - - - - - - -

Note: In logic, the biconditional is a connective of two statements that taken together is true if and only if both statements have the same logical truth value. In the above truth table p "if and only" q is true when both p and q have the same truth value. When both p and q are true the biconditional is true. When both p and q are false the biconditional is also true. When both p and q have different truth values then such a biconditional is false.

- - - - - - - - - - - -

EXERCISE 4

1. Is the following a biconditional statement?
   If I do not eat then I will be hungry.

2. What is a biconditional statement?

3. When is a biconditional statement true?

4. Explain the difference between a conditional statement and a biconditional statement?

5. Give two examples of a biconditional statement?

ANSWERS FOR EXERCISE 4

1. No. This is an "if-then" conditional statement. The "if "part is a hypothesis and the then part is a conclusion. The "if-then" conditional statement is false when the "if" part is true and the "then" part is false or when the "if" part is false and the then part is true.

2. A biconditional statement has an operator connecting two statements. The operator is s logical connective designated by a double arrow $\leftrightarrow$ and it means "if and only if."

3. A biconditional statement is true only when both statements have the same truth value, that is, when both statements are true or when both statements are false.

4. A biconditional statement is true only when both parts have the same truth value. A conditional statement is false only when the "if" part is true and the "then" part is false

5. I will get an "A" grade if and only if I get an "A" on the final exam.

A four-sided figure is a square if and only if it has four sides of equal length.

8. By means of a truth table we can show that the statement $p \rightarrow q$ is equivalent to its <u>contrapositive</u> $(\sim q) \rightarrow (\sim p)$.

p	q	$p \rightarrow q$	$\sim q$	$\sim p$	$(\sim q) \rightarrow (\sim p)$
T	T	T	F	F	T
T	F	F	T	F	F
F	T	T	F	T	T
F	F	T	T	T	T

Note: The contrapositive of a conditional statement is formed by flipping the statements and negating each one. Thus we make the contrapositive of "If p then q" rendered as $p \rightarrow q$ by flipping it and negating each part to read: "If not q then not p." rendered as $\sim q \rightarrow \sim p$ The contrapositive of "If it is a square then it has four equal sides" is "If it does not have four equal sides then it is not a square." Both statements are equivalent. The conditional is equivalent to its contrapositive.

EXERCISE 5

1. What is the contrapositive of $m \rightarrow n$?

2. Explain what is the contrapositive of a conditional statement.

3. Give an example of a true conditional statement and state its contrapositive.

4. Are the two statements you gave in question 3 both true. If so then explain why?

5. True or False: The contrapositive of a true conditional statement is also true.

6. True or False: The contrapositive of a false conditional statement is also false.

ANSWERS FOR EXERCISE 5

1. $\sim n \rightarrow \sim m$

2. The contrapositive of a conditional statement is formed by flipping the statements and negating each one. Thus we make the contrapositive of "If p then q" rendered as $p \rightarrow q$ by flipping it and negating each part to read: "If not q then not p." rendered as $\sim q \rightarrow \sim p$

3. Conditional: If an object is an apple, then the object is a fruit. Contrapositive: If the object is not a fruit, then the object is not an apple.

4. Both the conditional and the contrapositive statements in answer 3 are true. It is true that an apple is a fruit. It is true that if an object is not a fruit then it can not be an apple (because an apple is a fruit).

5. True.

6. True

9. An argument in symbolic logic consists of a set of given statements which are called premises from which is derived another statement called the conclusion. An argument is valid if and only if the conjunction of the premises implies the conclusion as a tautology. A tautology exists when an argument is always true for any truth values of its substatements.

Here is a simple example of an argument where the given two premises produce a conclusion:

Let p = I wash the windows
Let q = I clean the floor

First premise: $p \rightarrow q$ If I wash the windows then I clean the floor.

Second premise: p: I wash the windows

Conclusion: q: Therefore I clean the floor.

The argument can also be represented as follows:

Premise: If p then q
Premise: p
Therefore: q

This argument can also be represented as follows:

$[(p \rightarrow q) \wedge p] \rightarrow q$

[(p implies q) and p] implies q

EXERCISE 6

QUESTION 1: Test the validity of the following argument by constructing a truth table and show that it is a tautology.

p: John is a sophomore. q: John is taking biology.
ARGUMENTS:
(a) Given: $p \longrightarrow q$: If John is a sophomore, then he is taking biology.
Given: p: John is a sophomore.
Conclusion: q: John is taking biology.

ANSWER TO QUESTION NO. 1:

This argument is valid if and only if the statement $[(p \longrightarrow q) \wedge p] \longrightarrow q$ is a tautology.

p	q	$p \longrightarrow q$	$(p \longrightarrow q) \wedge p$	$[(p \longrightarrow q) \wedge p] \longrightarrow q$
T	T	T	T	T
T	F	F	F	T
F	T	T	F	T
F	F	T	F	T

The above valid argument is an illustration of the law of detachment (also called modus ponens). This law is a basic rule of inference in logic.

Note: In logic, the **Law of Detachment** says that the two statements are true, namely (1) If, then and (2). We can derive a third statement that is always true and therefore forms an all true tautology as indicated in the above last column of the truth table.

QUESTION 2: Test the validity of the following argument by constructing a truth table to show if it is a tautology.

Given: $p \longrightarrow q$: If John is a sophomore, then he is taking biology.
Given: q: John is taking biology.
Conclusion: p: John is a sophomoɪe.

ANSWER TO QUESTION NO. 2:

This argument is not valid because the statement $[(p \rightarrow q) \wedge q] \rightarrow p$ is not a tautology.

p	q	$p \rightarrow q$	$(p \rightarrow q) \wedge q$	$[(p \rightarrow q) \wedge q] \rightarrow p$
T	T	T	T	T
T	F	F	F	T
F	T	T	T	F
F	F	T	F	T

Note: The last column of truth values has an F. Therefore, the statement is not a tautology and the argument is not valid.

- - - - - - - - - - - - -

CONCLUDING REMARKS:

Logic can be applied to the study of the natural sciences, mathematics, computer science, law and philosophy. Its origin dates back to the time of the ancient Greek philosophers, especially noteworthy is Aristotle. The study of logic has developed over the centuries into a formal course with applications in many areas of knowledge.

Info about the below recent published math titles is posted at *www.wrparks.com* The titles are available at *Amazon.com*

Boolean Algebra and Switching Circuits

Sets, Numbers and Flowcharts

Computer Number Bases

Computer Number Bases -- Printed Text with Audio CD

Introduction to Logic

Beginning Algebra

Introduction to Gambling Theory – Know the Odds

Letters to a Young Math Teacher

Program Your Calculator

More books published by William R. Parks. Available at Amazon.com

Handbook for Piano Practice

Piano Practice for the Advancing Student

A Franciscan Odyssey

Windows to Heaven – Icon Painting

The Joyful Cook's Guide to Heavenly Greek Cuisine

The Nature Watch Collection Book One

The Nature Watch Collection Book Two

Animal Tails

The Calico Caterpillar

Peppin Puffin to the Rescue

Choices

Time to Fly

Jonah the Reluctant Prophet

Earth, God's Garden

Made in the USA
Monee, IL
04 November 2026

67377186R00015